26th Topical Conference on Refinery Processing 2023

Topical Conference at the 2023 AIChE Spring Meeting and 19th Global Congress on Process Safety

Houston, Texas, USA
12-16 March 2023

ISBN: 978-1-7138-7932-9

Printed from e-media with permission by:

Curran Associates, Inc.
57 Morehouse Lane
Red Hook, NY 12571

Some format issues inherent in the e-media version may also appear in this print version.

Printed with permission by Curran Associates, Inc. (2023)

For permission requests, please contact AIChE
at the address below.

AIChE
120 Wall Street, FL 23
New York, NY 10005-4020

Phone: (800) 242-4363
Fax: (203) 775-5177

www.aiche.org

Additional copies of this publication are available from:

Curran Associates, Inc.
57 Morehouse Lane
Red Hook, NY 12571 USA
Phone: 845-758-0400
Fax: 845-758-2633
Email: curran@proceedings.com
Web: www.proceedings.com

TABLE OF CONTENTS

RENEWABLE FUELS I

RENEWABLE FUELS II

INDUSTRIAL WASTEWATER MANAGEMENT AND WATER REUSE II

MEETING NET ZERO TARGETS: DEVELOPMENTS IN SUSTAINABILITY AND EMISSION REDUCTION II

22a Renewable Methanol Conversion to Lower Emissions Fuels

Lei Zhang, Arsam Behkish, David Dankworth, Mark Deimund, Keith Kuechler, Michael Lanci, Jason McMullan, Kirtan Trivedi

Lower Emissions Fuels (LEF) will be required to meet the CO_2 emissions mitigation due to evolving regulations. Innovation, use of proven or scalable technology, and lower system cost will be key elements for potential solutions. LEF can be made using variety of technologies, with each technology potentially utilizing a different feed source. In this presentation, we will be discussing various LEF technologies for producing gasoline and sustainable aviation fuel (SAF) from renewable methanol. The opportunities such as tailoring yield of specific molecules and further reducing capital cost and CI in methanol-to-sustainable products will be elaborated on to provide insights on future R&D directions in commercialization of renewable MeOH conversion technologies.

Renewable MeOH, as a versatile and lower carbon intensity (CI) intermediate, is anticipated to be produced in greater volumes during the energy transition. Advanced methanol conversion technology is a key enabler to address the world's growing demand for a range of lower emission fuels and lower CI chemicals. The presentation will also discuss ExxonMobil's recent methanol conversion technology efforts that have been focused on developing and deploying fully integrated scalable solutions to sustainable products from renewable methanol as the feedstock.

22b Green Ammonia, a Pathway to Green Fuels?

Ryan Perry, Eduard Lindner

Ammonia is one of the largest chemical markets by volume and its impact on the environment has therefore entered the spotlight as the global community looks to mitigate climate change by limiting emissions. There are two broad routes to make ammonia production 'greener' and more sustainable:

- Green ammonia involves using renewable power to produce hydrogen via the electrolysis of water. This can then be reacted via conventional ammonia synthesis to produce ammonia
- Blue ammonia involves capturing the emissions from a conventional ammonia facility before sequestering them

The improved environmental impact of green ammonia and policy targeting climate change are driving project announcements and interest in its adoption for fuel applications. This adoption faces many challenges such as technology scale up, policy developments, and market pressure. The number of variables and the emerging status of green ammonia leads to a large uncertainty in market forecasting and as a result, NexantECA has developed scenarios based on varying assumptions for adoption.

Objectives

Review of green ammonia technology and status

Review of green ammonia capacity outlook

Review of green ammonia in the fuel markets

CDU RETROFIT DESIGN ALTERNATIVES THROUGH PROCESS INTEGRATION BOOST REFINERY SUSTAINABILITY

Claudio L. M. Kuboski, Danilo H. Biazi, Helton S. Maciel, PETROBRAS, Rio de Janeiro, RJ, Brazil

Abstract

In the last decades, many oil refineries have implemented retrofit projects to reduce energy consumption. Although such projects incurred in lower carbon dioxide emissions, the main drive related to their implementation was economic. However, environmental guidelines are driving companies to stamp sustainability concepts into their business, including planning their future for carbon neutrality. On its strategic plan, PETROBRAS aspires to achieve net zero emissions (scopes 1 and 2) within a timeframe compatible with the Paris agreement. In the refining segment, we expect the carbon intensity measured in 2015 to drop by 16% until 2025, and 30% until 2030.

In 2021, PETROBRAS defined a goal to position five of its refineries (about 1.2 MMbpd) among the best refiners in the world in energy efficiency and operational performance until 2025.

To achieve that, PETROBRAS created a corporate program called "RefTOP – World Class Refining", which has tracked dozens of opportunities regarding process units and power generation in the sites assessed.

This work will focus on the methodology used to identify energy savings on some of the Crude Distillation Units (CDUs) based on pinch technology concepts and process integration, as CDUs are among the top energy consumers in a refinery.

In addition, we will present three case studies that are part of the RefTOP program.

The first case study presents how some process scheme modifications favor thermodynamics and allow decreasing the unit energy consumption in 28.5%, with a relatively minor impact on the preheating train. The second case study will explore the thermal coupling between CVDU and FCCU through hot feeding. We will present the required changes in the FCCU process scheme and the benefits in terms of operational costs, such as the removal of the fired heater and additional steam generation. The last case study will discuss the usage of residual heat on CDU to generate a tempered water loop. This water loop will allow the integration of the CDU and C3 splitter unit and replace steam consumption on the C3 splitter, thus decreasing the total site steam demand by 55 t/h.

KEYWORDS: Crude distillation; process integration; Pinch Analysis; Refinery; Energy Efficiency; Climate Change; Carbon Neutrality; Net Zero Emissions

Introduction

In 2018, refineries accounted for about 1.3 Gt CO_2 emissions globally representing 4% of total inventory (Sunny et al., 2022). The world experienced a decline in fossil fuel consumption in 2020 because of the COVID-19 crisis. However, energy consumption rose again to previous levels. Although some scenarios forecast a drop in oil consumption, mainly associated with the net zero emissions context, the IEA does not conceive a phase-out (World Energy Outlook, 2022).

In the last recent years, the number of countries and organizations that stated some commitment to meeting the net zero targets have increased significantly. In mid-2022, the net zero targets announced by governments corresponded to at least 83% of Green House Gases (GHG) emissions (Net Zero Tracker, 2022). The oil and gas companies are accountable for most of the GHG emissions, so many of them also have published statements about meeting net zero targets. Oil companies have been implementing projects to improve energy efficiency since the oil crisis in the 1970s when operational costs skyrocketed. However, in several cases, these projects did not decrease GHG emissions, but allowed revamps for capacity increase. Since then, academia has developed a massive amount of research in process designs with the onset of Pinch Technology. Nowadays, experts do not expect the oil demand to increase. However, it is required to decrease GHG emissions to comply with ESG commitments, making companies shift their focus to energy efficiency projects.

PETROBRAS is a Brazilian company primarily engaged in the exploration, production, refining, energy generation, and trading. In 2021, the company declared the ambition to neutralize its operational emissions of GHG (scopes 1 and 2) within a timeframe compatible with the Paris agreement. In the refining segment, the target is to decrease GHG emissions intensity by 16% until 2025, and achieving 30% reduction until 2030, compared to 2015, as shown in Figure 1 (PETROBRAS, 2022).

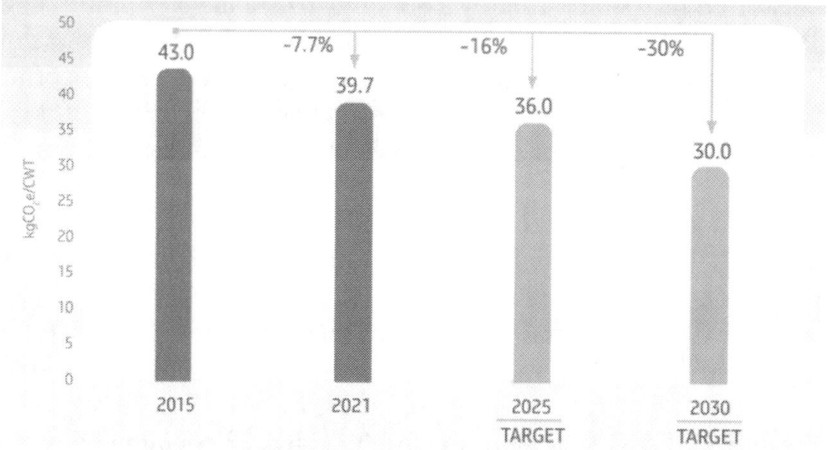

Figure 1: PETROBRAS Refining GHG emissions intensity

PETROBRAS created a corporate program called RefTOP—World Class Refining to address these goals. RefTOP intends to place PETROBRAS among the best oil refining companies in the world, by implementing actions to increase operational performance and energy efficiency. RefTOP program focused on five refineries (about 1.2 MMbpd), tracking almost 200 projects regarding Energy and Climate Performance. They oriented most of the projects toward decreasing primary energy consumption (natural gas, electricity and steam) and optimizing energy generation and distribution on refineries.

Initiatives to increase energy efficiency in refineries are among the most cost-effective tools to mitigate carbon footprint (Sunny *et al.*, 2022). Therefore, a process integration assessment can contribute to decrease GHG operations emissions effectively.

However, we have looked at process integration in a broader spectrum than the most frequent use of pinch technology. Besides opportunities for heat exchanger network retrofit, one must exercise a careful global review of process design. This work will present how to identify opportunities for decreasing energy consumption by modifying the configuration of a process unit, as well as exploring opportunities for integration between two interlinked process units. In addition, we have investigated opportunities to decrease the site steam demand through the thermal integration of process units.

Three case studies will exemplify part of the procedure used in the RefTOP program. First, we illustrate how to alter the process scheme to boost energy savings in a Crude Distillation Unit (CDU). In the second case, we explore process integration between CVDU and FCCU, through hot feeding. In the last evaluation, we focus on heat integration between two process units (CDU and C3 splitter) to decrease the site steam demand.

Methodology

RefTOP program portfolio is quite substantial and a considerable part of the benefits regards to better use of energy in process units. Therefore, we focused only on the method used to identify and evaluate potential process opportunities either through better heat integration within the processing unit or between process units.

In programs of this magnitude, the commitment and interaction between operational, R&D and process design teams are essential to accomplish good results. Besides understanding the global process flow scheme, one must know the details of the installation while considering design options.

We used pinch technology fundamentals to identify energy gaps in the original process unit design. Therefore, process integration expertise is crucial to develop a proper view of the opportunities.

The following sections describe the methodology used in this work. Although it does not present a novel approach, it was the basis for assessing significant results regarding energy efficiency.

Develop a base case scenario

A process simulation was the base to monitor operational parameters. We used the process simulation to guarantee that we were accurately addressing all the variables that affect energy efficiency. It should comply with typical operational parameters that are relevant to the objective of the work. The better the simulation adherence to historical data, the better its reliability.

If the engineer is primarily concerned in determining energy recovery potentials, the simulation does not need, *a priori*, to focus on a rigorous simulation of the preheating train. On the other hand, the preheating train simulation favors identifying flaws on the design and possible simple alternatives.

Data processing to generate the Grand Composite Curve (GCC) is challenging, and one must proceed carefully. One requires a DT min to create the GCC. We determined the dT min of the actual unit just by matching the GCC hot utility requirement with the actual furnace duty, ignoring any cross-pinch heat transfer.

To establish energy recovery targets, as an initial estimate, we assumed a dT min of 30°C and observed the reduction in furnace duty compared to the actual base case. We set the temperature at 30°C as an empirical trade-off parameter between additional energy recovery and requiring dramatic changes within the installed preheating train.

Evaluate the base case to find possible process improvements

One must observe three fundamental rules from pinch technology in process integration design:

1. Do not transfer heat across the pinch.
2. Do not place hot utility below the pinch.
3. Do not place cold utility above the pinch.

Following these rules can effectively drive the designer to pursue better heat recovery on its process scheme. As decreasing heating demand often directly correlates with GHG emission, it is essential to recognize process streams with potential to be a heat source above the pinch, which would incur lower hot utility demand. By accessing the GCC results, it is possible to observe the pinch temperature. As a result, one can derive an analysis of plausible improvements in process configuration.

We can exemplify the statements above on CDU, by looking at the pump-around (PA) circuits of the crude tower. Let us assume a tower configuration with three side streams and two PAs, one at the top and other close to the second side stream, as illustrated in Figure 2.

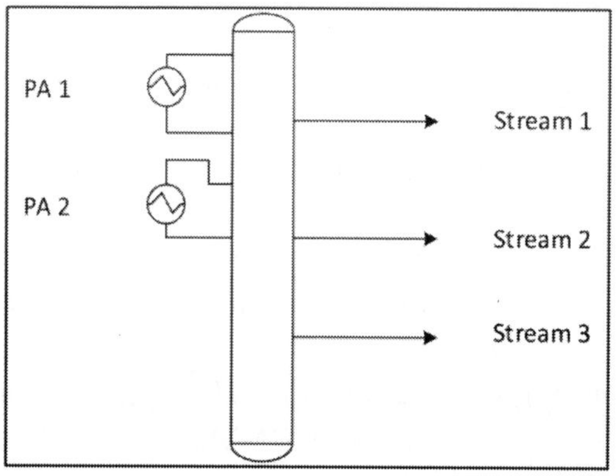

Figure 2: Crude Column configuration example

If the PA temperatures are lower than the pinch temperature, they are not contributing to decrease hot utility demand. However, if Stream 3 temperature is above the pinch, it is possible to create a new PA at this location, and the duty associated with this new PA above the pinch temperature will result in heating demand drop. It is also possible to manipulate the new PA flowrate to maximize energy transfer above the pinch.

The designer can also investigate decreasing the enthalpy required by cold streams above the pinch, which would also result in lower heating demand. For instance, one could consider installing a pre flash drum in a preheating train, lowering the heat demand. Another possible improvement concerns identifying hot utilities allocated bellow the pinch (e.g., tower reboilers), either using steam or flue gas as the hot energy source, and trying to eliminate this inefficiency.

On CDU, it is usual to send streams as feedstock to other process units, like FCCU or DCU. Most of the time, the stream is sent to tankage and then is routed to another processing unit. However, it might be worth investigating the effect of increasing the temperature of this inter-unit stream. If the temperature is below the CDU pinch, the only impact would be the decrease of cold utility requirements. Sending the stream at a temperature above the pinch in the downstream unit opens the scope for better global heat recovery.

Compare the effect of changes to the base case

After identifying an opportunity, the next step is check if the process change has a positive impact on heat recovery and to quantify it. For that, one can compare the energy targets at the GCC for the new alternative and the base case. If we find multiple opportunities, it is crucial to evaluate each one individually and then make an evaluation considering all of them together to determine how they interact with each other. This procedure allows the identification and classification of the opportunities in terms of energy benefits and capital costs, supporting the definition of the scope of the project.

Fine tune changes in the preheating train

After analyzing what modifications favor heat recovery on the assessed unit, one can look for improvements in the preheating train. The designer may choose a delta T min considered adequate for his purposes and seek improvements on the preheating train to achieve them. In CDU, defining a dT min means to define the temperature at the end of the preheating train as a target. Another alternative could be trying the most suitable modifications at the preheating train in terms of cost and feasibility and evaluating the impact concerning heating demand.

Check the alternative feasibility

The complexity associated with the modifications can be an issue. Retrofitting an existing preheating train is quite challenging because of many constraints, such as unit layout and available space to install new equipment, costs, hydraulics, materials, etc.

Also, when integrating different process units, a careful analysis of individual units' start-up and shutdown should be performed when evaluating the feasibility of the opportunity. One must look for a balance between appropriate heat recovery and complexity.

Case Study I

The first case addresses the usage of different process configuration schemes to improve crude and vacuum distillation units energy efficiency. We studied two different units, with slightly different process configurations regarding the heat exchanger network and distillation columns and capacity of 33000 m³/d each. Both have pre-flash tower for naphtha recovery, before the atmospheric furnace, and both have two pumparounds in the atmospheric tower, one on the top and one intermediate (refer to PA1 e PA2 on Figure 2). The main difference between the two process units is the usage of a reboiler in the pre-flash tower bottom. In terms of heat exchanger network, both units have three stages of exchange. One before the desalting unit, one before the pre-flash tower, and one before the atmospheric furnace. We will refer to them as the first, second, and third heat exchanger train. For the analysis, we followed the steps described in the methodology. First, in Figure 3, we show the grand composite curves for the base case for both units.

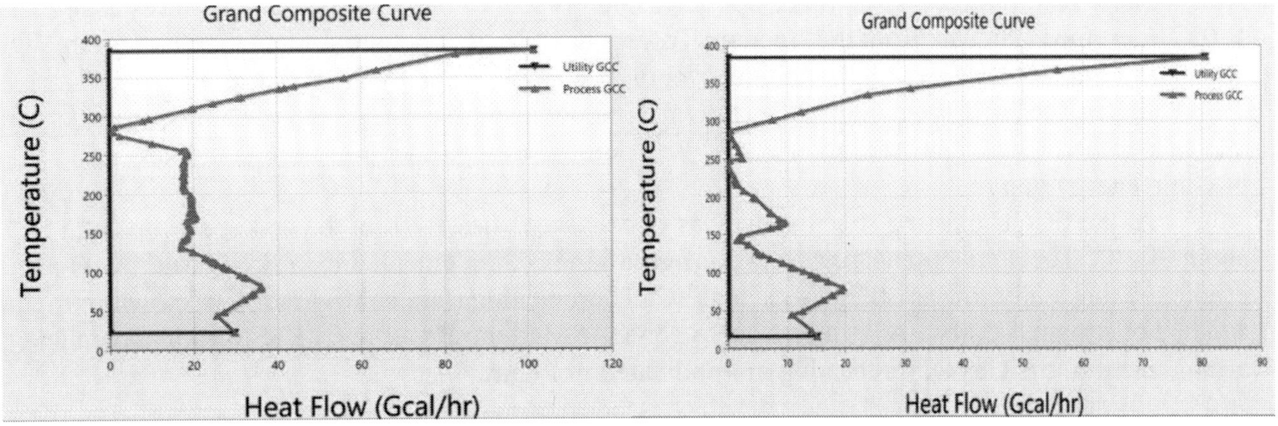

Figure 3: Grand composite curve for the base case: (a) CDU 1 and (b) CDU 2

For the CDU 1, GCC on the left, we observe a cold pinch temperature of 270 °C, which matches exactly the temperature for the reboiler in the pre-flash tower drum. This violates one of pinch fundamental rules, since part of the heat from the hot utilities is being exchange bellow the pinch temperature. Removing this reboiler must improve the energy efficiency of this unit.

In terms of process scheme, we can see two major opportunities. Recovering more naphtha from the pre-flash tower, hence reducing the amount of feedstock for the atmospheric furnace and creating an additional pumparound at the bottom of the atmospheric column. For the first one, we proposed an additional draw in the pre-flash tower in the intermediate section of the column. To define the rate of this draw we considered the minimum over flash flowrate to ensure a good operation of the tower. The latter is possible because in both cases the temperature of the pumparound is above the pinch. Since we only have one exchanger in the third heat exchanger network (using vacuum residue as a hot stream), we considered including a new heat exchanger with this new pumparound before this exchanger, following pinch rules.

We summarize the proposals for changes in process scheme as follows:

- A new draw from the pre-flash tower
- Relocating the pre-flash furnace reboiler
- Creating a new pumparound in the atmospheric column

In Table 1, we show the results obtained after the modifications in terms of the temperature before the atmospheric furnace and the heat demand in this furnace.

Table 1 – Results from case study I

	CVDU1 (before)	CVDU1 (After)	CVDU2 (before)	CVDU2 (after)
Temperature (°C)	**268**	**292**	**266**	**290**
Heat demand (MW)	**119**	**86**	**111**	**88**

Case Study II

Although this process integration is not a novelty, the results of this case were particularly interesting. In a Petrobras refinery, the process scheme is comprised of three Atmospheric Distillation Units and only two Vacuum Distillation Units. The VDUs are not thermally integrated to any of the Atmospheric Distillation Units. Atmospheric Residue is sent to the VDUs at temperatures as high as 180°C, therefore Gas Oils from the Vacuum Towers have limited potential to recover energy in VDU preheating train. Heavy Gas Oil is sent to air-coolers at approximately 200°C.

In this scenario, we conducted a study to find out a solution for a better heat recovery between VDUs and the FCCU. Hot feeding of FCCU was pointed by a partner consultant company as a possible energy gap.

At the FCCU existing scheme, the Gas Oil charge is heated from the tankage temperature up to about 300°C. The actual preheating train configuration is complemented by a fired heater, with an efficiency unusually low as 30%. The actual FCCU preheating train configuration is shown in Figure 4. The hot stream "Slurry PA" is also used to generate Medium Pressure (MP) Steam in a parallel arrangement to the Gas Oil preheating, not indicated in Figure 4.

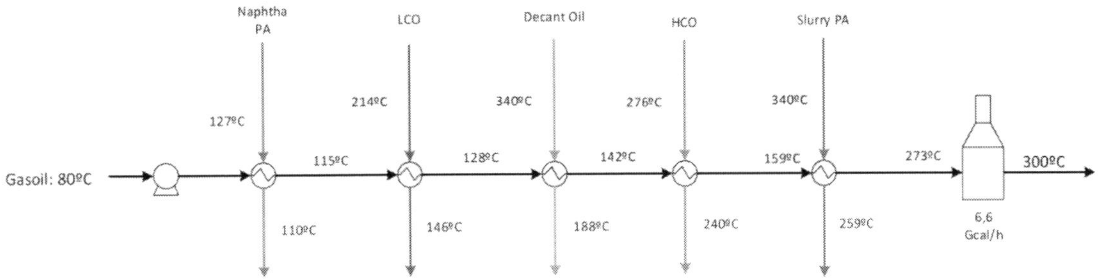

Figure 4: Base Case preheating train

Once the base case for the FCCU is built, some insights arise by analyzing the GCC. The GCC presented in Figure 5 matches the base case process furnace consumption (6,6 Gcal/h) at dT min of 30°C. The actual 44 ton/h MP steam generation is also included in the GCC. The pinch temperature observed is 111°C.

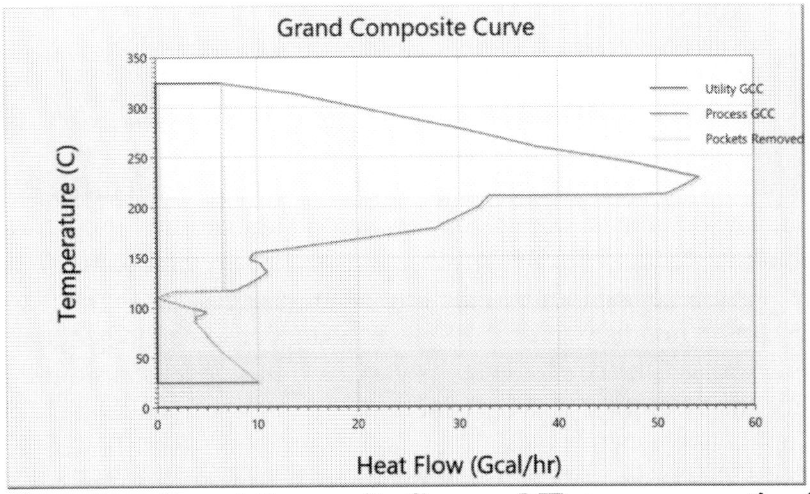

Figure 5: Base Case Grand Composite Curve – MP steam generation included.

It is important to understand the impact of MP steam generation on the base case. To do this, we have simply omitted this stream in the GCC, obtaining the curve presented in Figure 6.

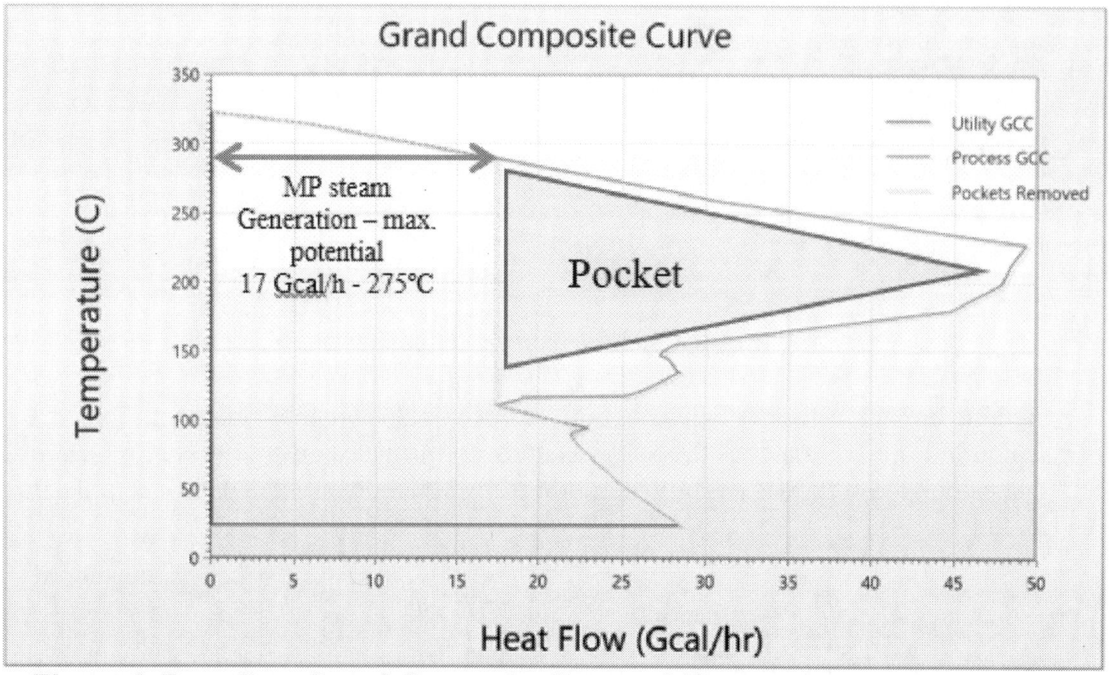

Figure 6: Base Case Grand Composite Curve – MP steam generation excluded

Just by this evaluation, it is possible to visualize and understand many aspects of the actual process integration. The theoretical MP steam generation obtained by Figure 6 is 17 Gcal/h, which is equivalent to 30 ton/h of steam (from BFW to superheated steam temperature). In the actual FCCU, if we want to reduce fuel gas consumption in the furnace, it is necessary to reduce MP steam generation from 44 ton/h to 30 ton/h. In other words, fuel gas is being consumed to generate more MP steam within FCCU. However, as furnace efficiency is very low, this operation should be avoided.

Putting aside the theoretical concepts and looking into the installed heat exchangers, we see that the FCCU operation do not switch between fuel gas consumption and MP steam generation deliberately. Actually, this is mainly a result of the heat exchanger performance at the Slurry Pumparound circuit. Slurry PA has a huge energy requirement at high temperature level (63 Gcal/h from 340°C to 240°C). In this FCCU, the Slurry PA is sent in parallel to three different services: preheating train (last service before furnace), MP steam generation and Debutanizer Reboiler. The heat exchangers in the preheating train do not have adequate excess of area/spare equipment. When they start to foul, to comply with the Main Fractionator energy requirements, it is necessary to increase MP steam generation. Yet, more energy must be added in the furnace to compensate degradation of the preheating train exchangers.

The impact of FCCU hot feeding was then assessed initially on the GCC, without including the MP steam generation. Figure 7 and Figure 8 presents how the GCC changes by setting FCCU feed temperature at 96°C and 200°C, respectively. By raising feed temperature from 80°C (Figure 6) to 96°C (Figure 7), it is possible to observe that the only difference between these cases is an increase in cold utility requirement, i.e., we have just shifted cold utility consumption from VDU to FCCU. But as we set feed temperature to 200°C (Figure 8), it is possible to observe that the potential to MP steam generation increased from 17 Gcal/h to an impressive value of 38 Gcal/h, maintaining the same cold utility requirement as 96°C feeding.

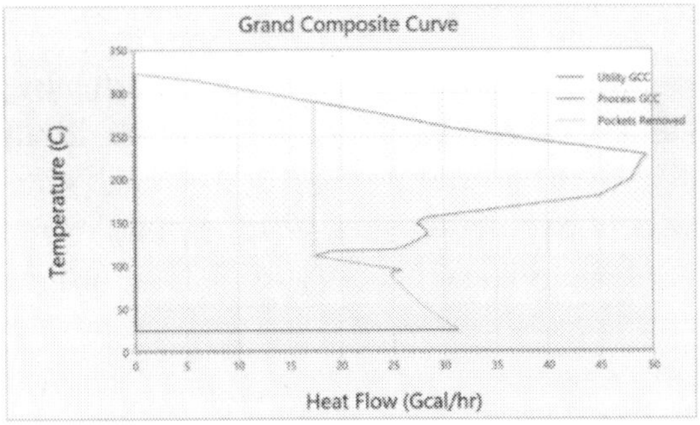

Figure 7: FCCU feed at 96°C.

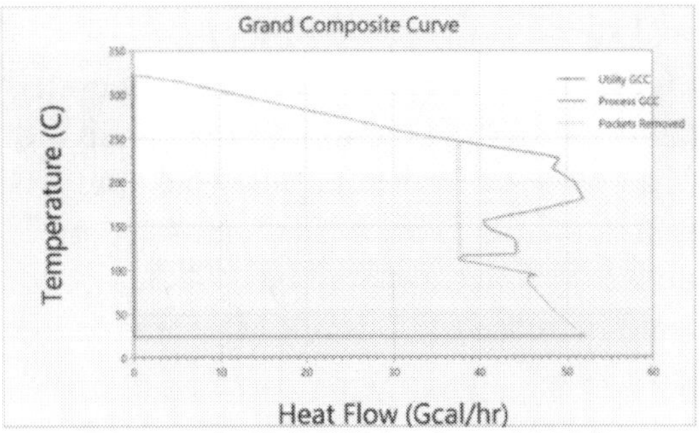

Figure 8: FCCU feed at 200°C.

If MP Steam generation is included in the problem at its maximum potential, a second cold stream pinch is created at 96°C. Therefore, any decrease on Gas Oil heating demand above 96°C contribute to an equal increase on steam generation amount.

Comparing the results of FCCU feed from 80°C to 200°C, the potential for MP steam generation increased from 30 ton/h to 67 ton/h. The drawback is additional 3 Gcal/h in cold utility requirement in FCCU.

This theoretical pinch background provided very useful information: it is possible to minimize fuel gas consumption in the low efficiency furnace, and it is also possible to generate additional MP steam by increasing FCCU feed temperature to 200°C. The refinery utilities experts indicated that the additional MP steam generation could provide additional capital savings by reduction in fuel gas consumption at Steam Boilers.

The next task is to evaluate which modifications are necessary to achieve these targets. Feeding FCCU at 200°C require some changes in preheating train. The first services should be relocated due to impossible temperature approaches, as presented in Figure 9.

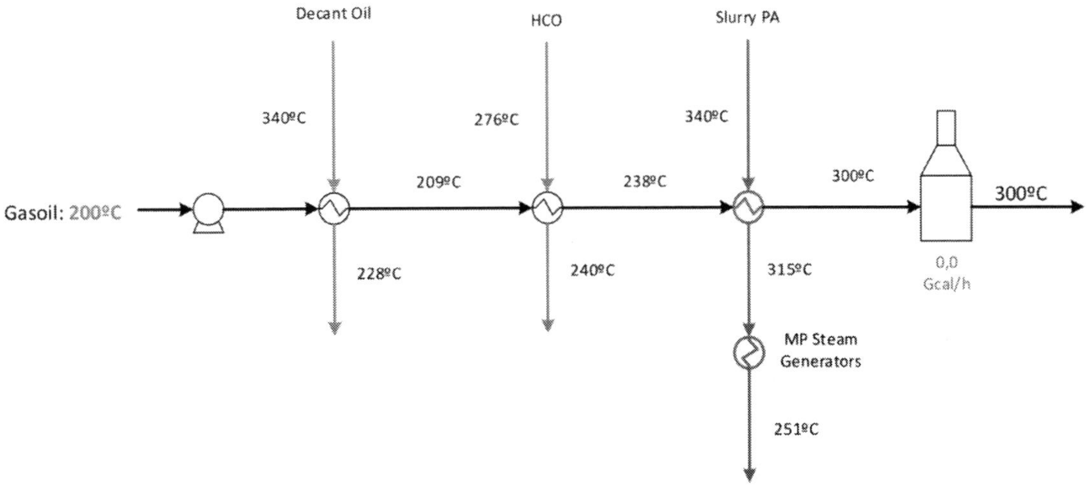

Figure 9: Proposed preheating train

Slurry PA circuit was carefully redesigned to allow for maximum heat recovery in preheating train and provide flexibility to maintenance. The MP steam generators were relocated in series with preheating train exchanger to allow higher velocities on each service. A new exchanger in the preheating train is proposed to guarantee that no fuel gas will be necessary, even when one of these heat exchangers is out for cleaning. A new MP steam generator was included to comply with the additional steam flowrate of 67 ton/h.

Other modifications were recommended to reutilize Naphtha PA and LCO preheating train exchangers along the Fractionation Area, avoiding that such energy would be lost to cold utilities. They will not be discussed in this work, but basically, we tried to minimize cross-pinch energy transfer. The whole project is summarized by 3 new heat exchanger shell, 2 tube bundles replacement, and pipe work. The estimated benefits are reduction of the released energy in FCCU furnace by 23,7 Gcal/h and increase of 23 ton/h MP steam generation.

Case Study III

The third case study regards a 40,000 m^3/d (252 Mbpd) CVDU. The unit main equipment comprises a pre-flash tower, originally associated to a reboiler furnace, a crude tower with 3 side streams and 3 pumparounds located just below each side draw, and a vacuum tower, with two side streams and pumparounds.

As part of the RefTOP program, this unit was reviewed under energy efficiency standpoint and some modifications, like those presented on Case Study I, were considered. As result, the pre flash furnace was relocated to operate as crude furnace, and the preheating train temperature achieved at the hot end was more than 285C. The proposed changes on the process scheme and on the preheating train configuration for this case will not be discussed, as the focus of this case study is about exploring one step further on heat integration.

At this moment, the focus of the work changed to identify opportunities to decrease total site steam demand and the main process consumers were identified.

In this context, it was observed that a C3 splitter unit required about 55 t/h of Low-Pressure Steam (LPS) to use as heat source for reboiling equipment (32.5 MW). The C3 splitter is composed by a deethanizer and a C3 splitter column. Both columns operate at low temperature at the reboiler section, as it can be seen on Figure 10.

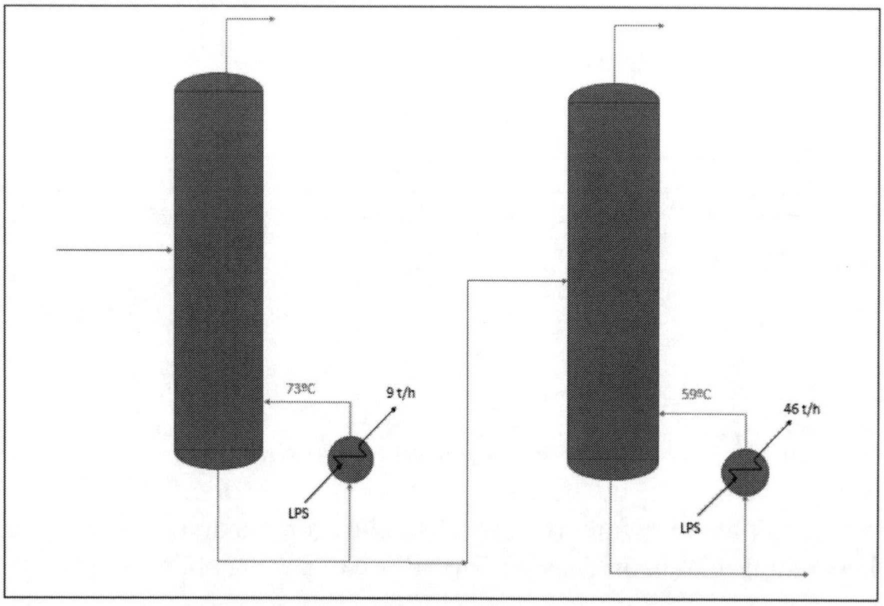

Figure 10: C3 splitter original design data

At this refinery, the C3 splitter unit is in the CDU's neighborhood. Therefore, it was quite convenient to investigate if heat integration between both units was feasible, thus minimizing pipework. Considering that the temperature level required by the C3 splitter reboilers is not that high, the energy that was being rejected to coolers on CDU could be used as an alternative to replace the LPS consumption, with no penalty to CDU preheating train performance.

The CDU original design included a tempered water loop used to cool appropriately heavy products, such as fuel oil and heavy vacuum gas oil (HVGO). The fuel oil stream is produced in specific campaigns, so the HVGO is the only stream rejecting about 11.9 MW of heat to the tempered water system regularly. The heat absorbed by the water in this system is released to the atmosphere through air-coolers. The tempered water loop runs at temperatures above 74°C. Therefore, a revamp for the tempered water loop could generate a hot source that would meet the C3 splitter unit energy requirements.

In order to avoid impact on the CDU preheating train, a search for process streams discharging heat to cooling equipment at temperatures above the required by the tempered water loop was carried out. Table 2 presents the process streams information used to provide additional heat, by the installation of new heat exchangers, to the revamped tempered water loop.

Table 2. CDU stream data for revamped tempered water loop

Stream	T in (°C)	T out (°C)	Enthalpy (MW)
A	141	91	11.4
B	126	86	5.0
C	193	91	4.2

Figure 11 illustrates the changes on the tempered water loop. The existing facilities and equipment are shown in black color, while the changes are highlighted in green color. It was required to double the system flowrate by adding another pump and distribute it between the new services. The water is then heated by the CDU hot streams available and is routed to the C3 Splitter unit, where is used as heating source for new reboilers, being cooled until 75°C. Finally, it returns to close the loop into CDU. The energy supplied by the old exchanger with HVGO and the new exchangers match the needs of the reboilers at the C3 splitter unit. If C3 splitter energy demand becomes lower than CDU supply, it is possible to release the surplus to the existing air cooler. On the other hand, if CDU energy supply is lower than C3 splitter unit demand, the old reboilers (using LPS) may complement the energy requirements.

Once the LPS demand is drastically decreased, it is necessary to implement additional adjustments on the power generation of the site. If the process demands less LPS, it is required to replace the driver of some pumps from turbine to electric motor, incurring in lower expansion of MPS to LPS. The same happens for equipment that expand HPS (consistent with the steam produced by the boilers) to MPS. By making these adjustments, the reduction on LPS results on a decrease on the steam produced in the boilers, incurring in a decrease of 83,000 t CO_2 equivalent per year, even considering the additional power importation for the electric motors.

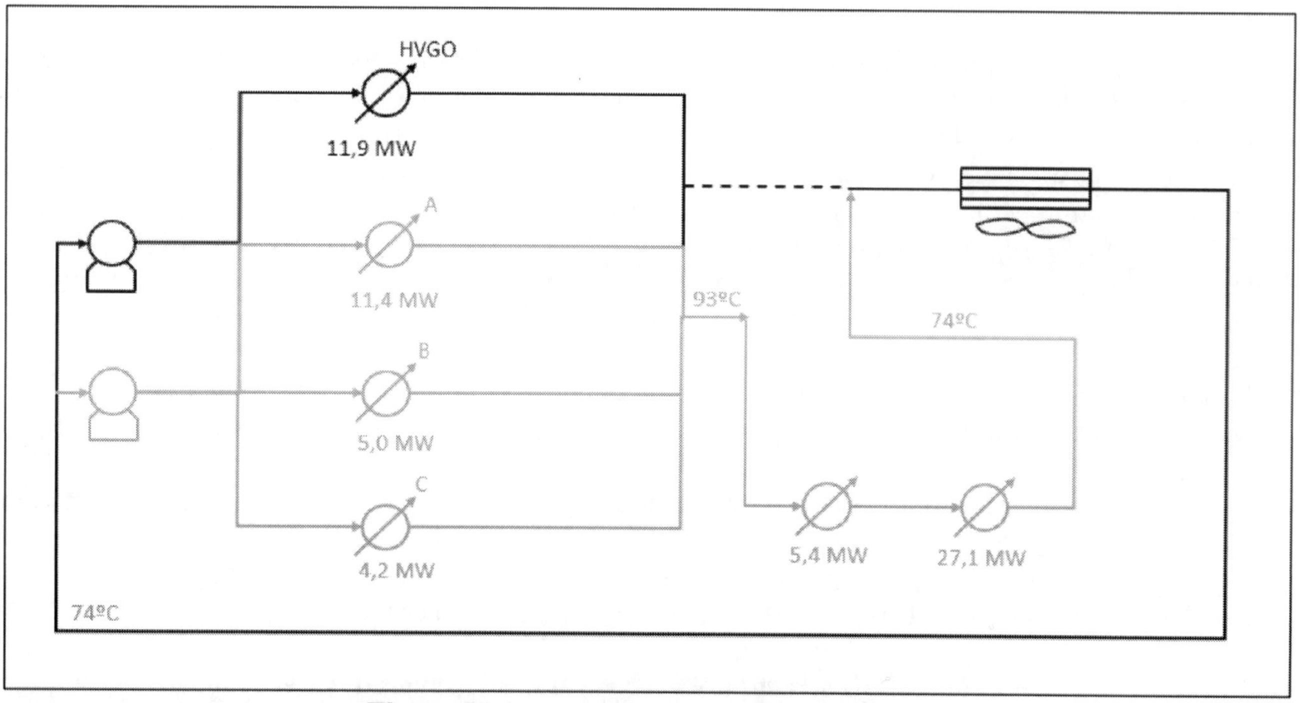

Figure 11: revamped tempered water loop

Another way of evaluating the impact of the project is by measuring the decrease on the Solomon's Energy Intensity Index (EII), usually considered to assess the energy efficiency on refineries. The benefit associated to this opportunity was identified as a decrease of 2.4 pts on Solomon's EII for the local site. Because of such impact on energy efficiency, this was considered one of the top projects on RefTOP program portfolio.

Conclusion

Many oil companies are changing their strategic plan to account for more efficiency, hence less polluting operations. In that regard, PETROBRAS released a corporate program focused on this subject called RefTOP. In this work, we discussed a methodology to evaluate existing refining units in terms of energy efficiency. We based our methodology on the pinch technology fundamental rules, often used to design new process units. We adapted this procedure for existing units in which several constraints are in play, such as space for new equipment. We presented three case studies, in which we adopted different solutions, conceptually speaking. The first one focused on changing process scheme to gain efficiency, the second focused on integration between process units, and last one addressed the issue of using residual heat as a hot heat source in other process units. We showed that minor changes might greatly improve energy efficiency and that each unit must be looked individually to ensure an optimal solution.

References

1. Sunny, Nixon, et al. "A Pathway Towards Net-Zero Emissions in Oil Refineries." *Frontiers in Chemical Engineering* 4 (2022): 4.

2. IEA. "World Energy Outlook 2022." (2022).
3. “Net Zero Stocktake 2022.” *Net Zero Tracker*, zerotracker.net/analysis/net-zero-stocktake-2022.
4. PETROBRAS. “Climate change supplement” (2022).

108a Key Considerations for the Selection and Design of Fixed Bed Axial Flow Vessel Internals

Eugene Kuchta

Key Considerations for the Selection and Design of Fixed Bed Axial Flow Vessel Internals

When designing Vessel Internals for a Fixed Bed Axial Flow process, the initial capital cost of a particular Vessel Internal should be balanced with its durability and its impact on the installation, maintenance, and turnaround schedules.

Compatibility with the process gas or liquid composition will determine the material of construction choices. Regenerative and non-Regenerative processes as well as piping arrangement will influence the type, design, and material of construction.

Inlet and outlet flow distribution, end-of-run pressure drop, vessel diameter, and adsorbent/catalyst particle size(s) along with total weight also need careful consideration when selecting and designing Vessel Internals.

This presentation will review and compare all of the above stated considerations when selecting and designing Vessel Internals.

108b Advances in Coker Unit Technology(Maximizing Refinery Petrochemicals, Minimizing CAPEX and Reducing Environmental Impact)

Gary Martin

Advances in distillation technology are leading to changes in coker unit fractionation schemes. These changes include incorporating revised or new fractionation to fix operating problems as well as improving molecular management to complete reconfiguration of new coker gas plant designs. An improved processing scheme utilizing dividing wall column (DWC) technology decreases plot space requirements, reduces capital investment costs, improves energy efficiency, and enables improved product recovery/separation. DWCs not only provide processing plants with lower CAPEX to meet a plant's processing objectives but they also provide improvement regarding environmental impact. These are advancements that make sense for the business world. That is, changes that improve business economics while at the same time do more with less and benefit the environment.

108c Equivalent Residence Time (ERT) As a Potential Correlation Parameter for Bitumen Visbreaking

Jose Angel Beleno Alcazar, Florian F. Schoeggl, Harvey W. Yarranton

The viscosity and density of bitumen must be reduced prior to transport in order to meet pipeline specifications. Typically, a diluent is added to do so, but this method reduces the capacity of the pipeline for bitumen. One approach to reduce the required amount of diluent is visbreaking. Visbreaking is a mild thermal cracking process that reduces the oil viscosity significantly and moderately reduces its density. The higher the severity (temperature and residence time), the greater the viscosity reduction. However, the practical extent of viscosity reduction is limited by the final product stability; that is, its tendency to precipitate asphaltenes or form coke.

To design a visbreaking process, it is desirable to predict the product properties and stability as a function of the reaction temperature and residence time. Visbreaking conversion increases as the temperature and residence time increase and both must be considered when attempting to predict the product properties. ERT is a measure of severity that combines the contributions of temperature and residence time into a single expression, given by (Yan, 1990):

where is the residence time in s, is the activation energy (209,500 J/mol), is the ideal gas constant in J/mol·K, is the reaction temperature in K, and is the reaction temperature reference (711.15 K). The objective of this study is to determine if conversion, product properties and stability follow consistent trends with ERT and, if so, to establish correlations of properties to ERT.

An Athabasca bitumen was visbroken in a continuous induction-heated reactor at constant pressure and over a range of severities. The following measurements were performed on visbroken products at different combinations of temperature and residence time: SimDist, C_7-asphaltene content, toluene insoluble (TI) content, gas yield, viscosity, and density. The distillate content and conversion were determined from the SimDist. The product stability was assessed in terms of the wt% of *n*-heptane required to cause asphaltene precipitation (onset).

As ERT increased, conversion was found to increase monotonically but in three distinct zones. In the first zone (ERT up to 1400 s), conversion increased significantly and almost linearly to 38%. The distillable fraction (Tb<367) increased from 17 to 36 wt%, the asphaltene content increased from 12.6 to 15.0 wt%, and there was no change in the TI content. The composition changes are consistent with scission reactions where the removed side chains contributed to the distillate content, and the partially stripped polynuclear aromatic cores added to the asphaltene content. In the second zone (1400 s < ERT < 3000 s), the conversion increased gradually from 36 to 42%. The distillable fraction increased from 36 to 39 wt%, the asphaltene content increased from 15.0 to 15.9 wt%, and the TI content increased to 0.06%. The small increase in distillates suggests that the scission reactions almost ceased while the increase in the TI content suggests that condensation reactions began. In the third zone (ERT > 3000 s), the conversion increased more steeply reaching 46.1% as the ERT increased to 3,200 s. The distillable fraction increased only slightly to 40.9%, the asphaltene content increased to 16.1% and the TI content increased to 0.08%. It appears that condensation reactions dominated in this zone. The gas yield was below 1.5% in all cases.

The changes in properties also follow distinct trends in each ERT zone. In the first zone the viscosity decreased from 816,000 to 1500 mPa·s, the density decreased from 1,012 to 993 kg/m³, and the onset decreased from 59 to 20 wt% *n*-heptane. These changes are consistent with the significant increase in distillate content. Changes in the asphaltene properties have also been shown to contribute to the decrease in density and viscosity (Marquez et al., 2020). In the second zone the viscosity decreased to 850 mPa·s, the density decreased to 989 kg/m^3, and the onset decreased to 19%, consistent with the small generation of distillables, asphaltenes, and toluene insolubles. In the third zone, the viscosity decreased to 460 mPa·s, the density increased with respect to the previous zone to 989.5 kg/m^3, and the onset decreased to 7%. These changes are consistent with condensation reactions dominating.

Similar property changes were obtained at same ERT even with different combinations of time and temperature. Hence, ERT is a sufficient parameter to correlate conversion, product composition, and properties, at least for one bitumen and a specific reactor. Preliminary correlations are proposed for conversion and product density, viscosity, and stability as a function of ERT. It remains to test the correlations on other feeds and reactors.

References

Marquez, A., Schoeggl, F. F., Taylor, S. D., Hay, G., and Yarranton, H. W. (2020). Viscosity of characterized visbroken heavy oils. *Fuel*, *271*(January), 117606. https://doi.org/10.1016/j.fuel.2020.117606

Yan, T. Y. (1990). Characterization of visbreaker feeds. *Fuel*, *69*(8), 1062–1064. https://doi.org/10.1016/0016-2361(90)90021-H

108d Hybrid AI for Industrial Thermal Systems

Francesco Coletti

Artificial Intelligence (AI) has disrupted several industries and for very good reasons. AI algorithms, combined with large computational power and scalability of the cloud are capable of finding non-obvious correlations in a large amount of data, predict future trends and help optimise various types of operations. They can be very efficient (i.e. can process a vast amount of data relatively quickly) and easy to apply when physics are not known. However, when deployed to the process industry, they have specific limitations including the fact that i) they rely on large quantities of data and, crucially, on their quality and ii) their extrapolation abilities are limited.

In this presentation we will discuss how deterministic models can be combined with AI ones into an Hybrid-AI approach and how this approach is currently being used in refineries to overcome the traditional limitations of AI methods. Specific case studies for various systems, including crude distillation units will be shown.

122b Effective Decolorization of Wastewater By Chemically Activated Biochar

Payal Maiti, Bhim Charan Meikap

Water is the most valuable gift from nature on the earth for sustaining our healthy life. However, the modernization of industries caused significant environmental damage, mainly through the discharge of industrial wastewater, which severely affected the water resource and its purity. The textile industries are identified as the most significant polluting source that continuously harms the environment out of all the industrial sectors. As consumer demand escalates for textile products, the textile industries and associated wastewater have increased proportionally. Each year, the global yield of about 0.7 million tons of dye in 10,000 different variations, and 2-50% of that are dumped as untreated polluted water. This dye contained waste effluent through industrial application discharged into the environment, which has a negative impact on flora, fauna, and human beings. It involves a decrease in sunlight penetration on the water surface, which interrupt the photosynthesis activity of aquatic plants. Furthermore, it is also responsible for various issues, such as cancer, genetic mutation, nausea, vomiting, and many more. In particular, Malachite Green (MG), a typical cationic dye, has been widely utilized in industries for dyeing silk, cotton, wool, leather, paper, and jute. It is also used in fish farming to cure fungal and bacterial infections. However, the residual dyes present in the effluent, even in small amount, can harm the ecosystem and causes health risks to society. MG affects the reproductive system and immunity due to its properties of genotoxicity and carcinogenic. Various techniques have been adopted to address these problems and eliminate the harmful effects. These include solvent extraction, co-precipitation, flocculation, ion exchange, adsorption, and membrane separation. Different studies from past decades have indicated that the adsorption process is one of the most effective techniques to treat dye molecules from polluted water due to its simple operation with higher removal efficacy, availability of the adsorbent, and cost-effectiveness. This study aimed to synthesize the bio-adsorbent for MG dye treatment. For the fabrication of the adsorbent, carpentry waste sawdust was employed as carbonaceous material, and for chemical activation, $ZnCl_2$ was utilized. Several analytical techniques were employed to study the physicochemical characteristics (i.e., XRD, FTIR, and FESEM/EDX). Moreover, a batch-scale decolorization experiment was carried out to study the effect of initial pH, adsorbent amount, and interaction time. The maximum removal of MG dye (99.03%) was achieved at a pH of 9.5. The experimental adsorption kinetic data of MG dye followed the pseudo-second-order model. Furthermore, the dye removal efficiency of the adsorbent was 80.86% obtained at its pH, which was increased by 1.5 times after chemical activation. Overall, these results implied the adsorption potential for successful uptake of effluent from textile industries where the released dye is of similar electropositive nature as that of MG dye.

122c Oil/Water Separation Using a Naturally Processed Demulsifier

Ahmad Adewunmi, Muhammad Shahzad Kamal, Syed Hussain, Shirish Patil, Afeez Gbadamosi, Dhafer Al-Shehri

The goal of this study is to investigate the potency of extracted corn husk particles as demulsifiers (emulsion breakers) for crude oil emulsions produced in the petroleum industry. Extracted husks from corn were milled into fine particles, characterized, and employed as demulsifiers for water removal from heavy crude oil emulsion produced during hydrocarbon exploitation. The emulsion was formed from crude oil and synthetic seawater. Various amounts of corn husk particles acting as demulsifiers were added into 10 mL of W/O emulsion in separate 15mL bottles tightly covered and agitated. Oil/water separation was monitored inside the oven. The operating temperature was set at 65 °C. The corn husk particles performed excellently in the separation of water from crude oil emulsions. The particles were able to settle at the water bottom without mixing with the oil phase. Optical morphology of produced emulsion before and after the addition of husk particles showed traces of oil sample are still left in the water. Rheological characterizations support the observation from the demulsification test. The mechanism governing the performance of the corn husks demulsifier for the emulsion separation was proposed. The outcome of this study reveals that corn husk particles are promising demulsifiers for water-oil separation in the petroleum industry and are capable of achieving approximately 99% demulsification efficiency. This is the first study to investigate the efficiency of corn husk particles as a demulsifier for heavy crude oil emulsion separation. The use of corn husk particles will save operational costs as the known commercial demulsifiers are very expensive.

93a Decarbonization of Refining Processes By Real-Time Monitoring and Knowledge-Based System for Mitigation of Petroleum Fouling

Chandrakant (CB) Panchal

Petroleum fouling is a major unsolved problem costing the industry in terms of productivity penalty, maintenance costs and higher energy usage. The recent trends in environmental concerns of green-house gas emissions, increased energy and maintenance costs, and impact on plant capacity and availability are governing motives for industrial interests in cost-effective and practical fouling mitigation in petroleum processing. The physical and chemical mechanisms associated with petroleum fouling are complex with interactive effects of precursors derived from blending of feed sources. The petroleum processing starts from crude distillation to down-stream processing to production of petrochemicals. Significant progress has been made; however, implementation of monitoring and root-cause analysis (RCA) based mitigation has been slow. However, with the initiative of decarbonization in the process industry effective mitigation of fouling is getting a renewed attention along with heat integration and utilization of waste heat. In this investigative assessment study, an overview of the past and recent case studies is evaluated with the focus on energy efficiency in petroleum processing and corresponding abatement of CO_2 emissions. An extensive Exxon Chemical 1979 study of Economic Penalties Associated with the Fouling of Refinery Heat Transfer Equipment provided a fundamental basis for further development of mitigation technologies. In a follow-up 2000 study by Argonne National Laboratory focused on the Effects of Mitigating Fouling on the Energy Efficiency of Crude Distillation Unit (CDU). Recently in 2020 AIChE Sprint meeting Monitoring-Based Fouling Mitigation in Crude Preheat Trains (PHT) Using Lab and Field Fouling Units focused integration of field fouling monitoring with simulation software of the system. Such as an integrated system provides a real-time knowledge-based monitoring for taking corrective actions to minimize energy and productivity penalties. Similar investigative studies have demonstrated that the effective mitigation of petroleum fouling would have a direct impact on energy efficiency and hence abatement of CO_2 emissions per unit petroleum products. In this assessment study a systematic roadmap is proposed to link the knowledge and real-time monitoring-based mitigation of petroleum fouling and energy consumption. It consists of three interlined knowledge-based system: a) real-time monitoring; b) information management and c) responses. This integrated knowledge-based system can be implemented to operating individual processing units (fired heaters and feed/effluent exchangers), subsystem (PHT) or process unit (CDU and hydrotreaters). It is crucial to accurately and quantitatively monitor the effectiveness of mitigation actions, including anti-foulant treatments. Such mitigation techniques might have shown the effectiveness in lab tests, but less effective in the process unit. Case studies have shown that the effectiveness of physical and chemical mitigation techniques is dependent on fluid dynamics that may not be simulated in lab tests. The knowledge-based system will overcome this limitation for qualifying cost-effective mitigation techniques. The resulting outcome the knowledge-based system has direct impact on the enhancing the energy efficiency in petroleum processing and hence abatement of CO_2 emissions that can justify the short-term as well as long-term corrective actions to be implemented.

93b A systematic approach to correlate fouling deposit analysis to root causes of heat exchanger fouling

Himanshu Joshi, Francesco Coletti

Fouling in heat exchangers is a cause of severe economic losses, increased CO_2 emissions and disruption to normal operations. In this paper, we will revisit in details the mechanism leading to fouling discuss the analytical techniques that can be used to unveil such mechanisms. Using practical examples we will show how basic and advanced measurements can be applied to diagnose fouling issues and identify root causes. This information can the be employed to find suitable mitigation methods.

93c Asphaltene Deposition in Vertical Open Flow

Maria Gomez Rodriguez, Florian F. Schoeggl, Harvey W. Yarranton

Asphaltenes can precipitate and form deposits in open flow in wellbores, pipelines, and oil and gas facilities. These deposits can lead to production losses and high treatment costs; hence, deposition models are used to predict the conditions at which deposition is a risk. Asphaltene deposition at near ambient temperatures, where asphaltenes are commonly considered to precipitate as glassy particles, involves the adhesion and accumulation of asphaltene molecules, particles, and aggregates at the surface of a pipe or vessel. Several models have been developed for asphaltene deposition at these conditions which account for asphaltene aggregate size and concentration, shear conditions, fluid properties, and surface properties [1-2]. However, asphaltene deposition at the higher temperatures encountered in deep, offshore production has not been rigorously investigated. At higher temperatures, asphaltenes separate from the oil as part of a heavy liquid phase (liquid droplets) rather than as glassy particles and may deposit differently.

In a recent study, an asphaltene deposition apparatus was commissioned to investigate asphaltene deposition in horizontal laminar flow in both the glassy particle and liquid droplet regimes [3]. The apparatus consists of two transfer vessels, a co-axial nozzle-like static mixer, and a capillary tube test section, all housed in an air bath. Metered positive displacement pumps are used to displace pre-diluted bitumen and *n*-heptane, respectively, from the transfer vessels to the test section. In a deposition experiment, the differential pressure across a capillary tube (an indicator of asphaltene deposition) is measured during the flow period. In addition, the capillary tube is removed after each experiment to measure the mass, solvent content, and location of the deposit.

In the previous study, asphaltene deposition in horizontal laminar flow was evaluated in a 1.7 mm ID capillary tube for mixtures of bitumen and *n*-heptane, with *n*-heptane contents ranging from 65 to 90 wt%, fluid flow rates from 2 and 4 cm^3/min, capillary tube lengths from 3 to 30 cm, and temperatures of 50, 90, and 130°C. It was found that, in the glassy particle regime (50 and 90°C), precipitated asphaltene particles formed a porous deposit near the inlet of the capillary tube with cycles of deposition and erosion occurring during the flow period. In the liquid droplet regime (130°C), asphaltene-rich heavy phase droplets settled and coalesced to form a continuous heavy phase layer leading to stratified flow with occasional temporary plugging.

The main objective of the current study is to determine if the deposition mechanisms change in vertical flow. The previously commissioned asphaltene deposition apparatus was modified to evaluate asphaltene deposition in laminar vertical flow in both the glassy particle regime (50°C) and the liquid droplet regime (130°C). Deposition (or multiphase flow) in a 1.7 mm ID capillary tube was assessed for a test fluid of bitumen diluted with *n*-heptane. Feed composition from 65 to 90 wt%, flow rates of 2, 4, and 8 cm^3/min, and tube lengths from 3 to 30 cm were investigated. The preliminary results were qualitatively similar to those for horizontal flow. In the glassy particle regime, the differential pressure cycled, consistent with periods of deposition and erosion. The mass of the deposit was small and its location varied in each experiment. In the liquid droplet regime, the pressure drop increased exponentially in the first 40 min until the heavy phase occupied approximately 80% of the capillary tube volume. The flow regime up to this point was interpreted as core-annular flow with the heavy phase coating the surface of the pipe and the light phase carrying entrained liquid droplets in the core. After 40 min, low amplitude cycles of increasing and decreasing pressure drop were observed, consistent with slug flow. In summary, as observed in horizontal flow, deposition occurred in the glassy particle regime while multiphase flow was observed in the liquid droplet regime.

References

1.Vargas, F.M., Creek, J.L., Chapman, W.G. On the Development of an Asphaltene Deposition Simulator. *Energy Fuels*, 24, 2010, 2294-2299.

2.Eskin, D., Ratulowski, J., Akbarzadeh, K., Pan, S. Modelling Asphaltene Deposition in Turbulent Pipeline Flows. *J. Chem. Eng.*, 89, 2011, 421-441.

3.Do, "Effect of Temperature on Asphaltene Deposition Mechanisms in Horizontal Flow," University of Calgary, 2021.

58a Renewable Diesel Production through Stand-alone and Co-hydrotreating of Catalytic Fast Pyrolysis Oil

Xiaolin Chen, Kristiina Iisa, Calvin Mukarakate, Kellene A. Orton, Michael B. Griffin

Catalytic Fast Pyrolysis (CFP) of biomass followed by hydrotreating can be used for the production of renewable transportation fuels, such as sustainable aviation fuel and renewable diesel, which can reduce the dependence on fossil fuels and help achieve greenhouse gas reduction goals. Stand-alone hydrotreating of CFP oil and co-hydrotreating of the CFP oil with straight-run diesel (SRD) showed potential to produce high-quality diesel products in this study. Stand-alone hydrotreating of CFP oil produced a diesel fraction with oxygen content < 0.1 wt.% and an indicated cetane number (ICN) of 45. Co-hydrotreating of CFP oil and SRD using a 1:4 volumetric ratio produced diesel products with 0.1 wt.% of oxygen and ICNs of 42~45. Compared to $CoMo/Al_2O_3$, $NiMo/Al_2O_3$ was a more attractive catalyst for co-hydrotreating of two streams, leading to a higher carbon efficiency, a higher selectivity of cycloalkanes, a higher ICN, and enhanced desulfurization. Compared to stand-alone hydrotreating of CFP oil, co-hydrotreating with SRD could be achieved at a less severe operating condition and it produced hydrotreated products with a higher carbon efficiency. Synergy between CFP oil and SRD during co-hydrotreating process was suggested by the better experimental results in comparison with theoretical interpolation results. Co-hydrotreating gave high biogenic carbon incorporations of 91%~97% as determined by C-14 analysis.

58b The integration of experimental hydroprocessing and FCC data with process (Aspen Plus) and refinery optimization (Aspen PIMS) models

Michael Talmadge, Nicholas Carlson, Abhijit Dutta, Michael B. Griffin, Kristiina Iisa, Calvin Mukarakate, Reinhard Seiser

The integration of experimental hydroprocessing and FCC data with process (Aspen Plus) and refinery optimization (Aspen PIMS) models

58c Co-processing of Bio-Oils in FCC Reactor

Reinhard Seiser, Jessica Olstad, Rebecca Jackson, Clark Yarbrough, Kim Magrini

NREL has co-processed two types of bio oils (CFP oil and stabilized FP oil) with VGO using two catalysts in a Davison Circulating Riser. We have analyzed products for biocarbon and find that relative biocarbon incorporation into the liquid product is increased when adding Johnson Matthey's CP758 catalyst to E-Cat.

58d Co-processing bio-oils with petroleum feeds to produce fuels

Calvin Mukarakate, Cody Wrasman, Michael Talmadge, John Kayser, Luke Tuxworth, Kimberly Magrini-Bair

Co-processing bio-oils with petroleum feeds in a fluid catalytic cracker (FCC) is a near-term approach for introducing renewable carbon into fuels to help reduce green-house-gas emissions in the transportation sector. It is well documented in literature that co-processing raw bio-oils with vacuum gas-oil (VGO) using standard equilibrium catalyst (E-cat) will reduce yields of desirable products (gasoline, light cycle oil (LCO), and LPG-olefins) due to increased production of dry gas, coke, and heavy cycle oil (HCO). This is not surprising because the Y-zeolite, an active component of E-cat, is not effective for catalytic upgrading of biomass derived molecules. In this work, we will use a different approach to enhance biogenic carbon incorporation in desirable products, gasoline, LCO and LPG-olefins. The results presented in this work were obtained using an ACE reactor (Kayser Technology Inc.) equipped with a dual feeding system for bio-oils and VGO. We will first present on the effect of co-processing catalytic pyrolysis-derived bio-oil (CFP-oil), an upgraded bio-oil with ~13wt% oxygen content, with VGO on FCC product yields. We will then report on the impact of modifying E-cat with additives, that targets, conversion of biomass derived molecules. The intended outcome is to provide guidance on how to enhance biogenic carbon incorporation in desirable fuel range molecules and LPG-olefins by making small modifications to the E-cat. The results from these studies will be used to inform economic models for co-processing bio-oils with VGO and highlight opportunities for accelerating the deployment of the co-processing technology.

75a Research on Sustainable Aviation Fuel at the National Renewable Energy Laboratory

Robert Baldwin

The US Department of Energy has issued the SAF Grand Challenge that sets volumetric goals for production of sustainable aviation fuel (SAF). This presentation will focus on research and development on SAF at the National Renewable Energy Laboratory. Novel pathways being developed from waste and renewable feedstocks – including CO_2 - will be highlighted along with information on economics and sustainability.

75b A Model-Informed Approach for Fixed Bed Catalyst Testing and Process Development at Pre-Pilot Scale

Bruce Adkins, Canan Karakaya

The traditional path for fixed bed catalytic process development is lab scale to bench, pilot, demonstration and finally commercial. Each step represents a substantial increase in time and expense, so there is great interest in streamlining the process.

Initial experiments using powdered catalysts are typically isothermal and gradientless, and produce essential fundamental information. With increasing scale, the following changes occur:

- Pelleted catalysts: catalysts used in commercial fixed bed reactors always take the form of "FFF" (fully formed and formulated) pellets. On one hand, FFF catalysts represent activity loss compared to catalyst powders on a reactor volume basis due to (1) bed voidage and pellet porosity, (2) additional components (binders, fillers etc) needed for physical properties, and (3) internal mass and heat transfer (MHT) limits. On the other hand, added components can impart very useful additional properties, like improved deactivation resistance and multifunctionality. Importantly, the differences between powdered and "FFF" catalysts are very complex and not easily modeled a priori, but must be addressed through testing and characterization.
- Bed gradients: as bed size increases, gradients in flow, concentration and temperature develop as a consequence of bed (and external) MHT limits. These also can have the effect of strongly reducing the volumetric activity.

The BETO (Bioenergy Technology Office) funded consortium known as CCPC (Consortium for Computational Physics and Chemistry) has developed a concept for using computational models in combination with testing of FFF catalysts at "pre-pilot" scale: i.e. the smallest scale capable of measuring the parameters necessary for accurate pellet, bed and reactor modeling. In doing so, we aspire to accelerate the traditional scaleup process and potentially reduce the need for lifetime testing.

In this presentation, we will share examples of this concept in theory and practice.

75c Co-Processing Fischer-Tropsch Wax and Vacuum Gas Oil in a Pilot-Scale FCC

Reinhard Seiser, Robert Baldwin, Jessica Olstad

The long-haul aviation industry presents a particularly difficult problem for decarbonization; switching to fuels with some or all of the carbon from renewable resources is seen as the only viable solution that is currently available. This paper will describe R&D on co-processing of Fischer-Tropsch wax with vacuum gas oil in a pilot-scale fluid catalytic cracking system (Davison Circulating Riser or DCR) along with hydrotreating for fuel finishing. The DCR has been 'tuned' to allow for more production of middle-distillates such as jet fuel. A feature of this paper is tracking of biogenic carbon from the feed to the final products.

75d Economic and Sustainability Assessment on Bio-based 2,3-BDO Separation Approaches for Sustainable Aviation Fuel Production

Eric Tan, Hakan Olcay, Jacob Dempsey, Jian Liu, Charlie Freeman, William L. Kubic Jr., Yuan Li

Sustainable aviation fuel (SAF) plays a critical role in aviation decarbonization. SAF can be derived from lignocellulosic biomass, such as corn stover, via 2,3-butanediol (BDO) intermediate. BDO undergoes downstream upgrading, including dehydration, oligomerization, and hydrotreating, to make the hydrocarbon blend stock like SAF.

Separating BDO from a fermentation broth is challenging. Water is more volatile than BDO, so energy consumption for ordinary distillation is prohibitively high. For BDO to be a feasible intermediate for sustainable biofuels such as SAF, the total energy usage for the BDO separation target was set to be no greater than 30% of its lower heating value (LHV).

We have developed and explored less energy intensive separation technologies for processing dilute fermentation BDO broth into suitable feed for downstream upgrading. The combined economic and sustainability assessment was performed to assess the feasibility of select cost-effective process designs and comparisons with baseline technology (i.e., cascade vacuum distillation).

137a Optimal Design of Liquid-Liquid Extraction for Wastewater Streams and Potential Application to Remove Emerging Contaminants

Donald Glatz, Brendan Cross

Abstract: Optimal Design of Liquid-Liquid Extraction for Waste Water Streams and Potential Application to Remove Emerging Contaminants

As water resources become scarce and the industry continues to move toward decarbonization, the chemical processing industry is challenged to find new and innovative ways to recycle its wastewater and further reduce energy consumption. This challenge is further complicated by increased regulatory compliance to both remove regulated contaminants to increasingly lower concentration levels as well as address emerging contaminants such as Per- and Polyfluoroalkyl Substances (PFAS). Given that emerging contaminants such as PFAS break down very slowly over time, making the removal of PFAS from our environment extremely difficult and energy intensive.

When assessing which technology is optimal to remove contaminants and recycle wastewater streams, available options for consideration are determined by whether streams contain organic chemicals that are either lower or higher boiling than water. Distillation or stripping is usually used to remove the lower boiling components, whereas liquid-liquid extraction is often used when the chemicals are either higher boiling than water or are non-volatile.

When it comes to addressing wastewater streams containing high boilers, a process widely used in the organic chemistry field is an extraction step allowing for counter-current contact with an immiscible solvent that is lower boiling than water followed by a distillation column to recover the solvent as an overhead stream for recycling back to the extraction column and a stripping column to strip and recovery the solvent from the aqueous raffinate stream.

However, for years a different technology has been employed by the mining and metals industry for the recovery and purification of non-volatile metals from aqueous streams. This technology has the potential for application in the chemical processing industry as a more energy-efficient process to treat wastewater streams containing high boilers. For example, extraction for the recovery of metals from dilute aqueous streams typically involves several extraction columns operating in series without the need for any distillation step, and as such these processes can often have lower operating costs than their counterparts in the organic chemical field. A high boiling Extractant is dissolved into a diluent, which is also higher boiling than water. Following the extraction step, the extracted metals are then re-extracted (or stripped) out of the "loaded solvent" using a second extraction step. The solvent used in this column is called a stripping solution which is often pH-adjusted water.

In order to provide a more efficient and economical solution to the decarbonization of the chemical processing industry while both addressing increasing water scarcity and future energy-intensive remediation challenges, such as mitigating emerging contaminants, the organic chemicals industry should learn from the metal extraction industry and consider applying multiple extraction columns in series without a distillation step as an alternative process improvement solution to address wastewater streams containing high boilers.

This paper will be presented using both types of processes; extraction followed by a distillation step and extraction using a high boiling solvent in multiple extraction columns. The selection of which one is best for a particular wastewater process requires an economic study for both methods.

By: Donald J. Glatz and Brendan Cross, Koch Modular Process Systems LLC, Paramus, New Jersey, USA

Author Bio

Don Glatz is the Manager of Extraction Technology at Koch Modular. Don's activities include the evaluation and optimization of extraction processes plus the scale-up and design of extraction systems. He has been working in this field for the past 30 years and has published a number of papers and articles covering this subject. Don holds a BS in chemical engineering from Rensselaer Polytechnic Institute and an MBA from Fairleigh Dickinson University.

Brendan Cross is a Principal Extraction Engineer at Koch Modular that works in the Extraction Technology Group. Brendan is responsible for liquid-liquid extraction application evaluation, extraction process development, and pilot test design, extraction column design, and commissioning & process startup. He has been with Koch Modular for 15 years and also has substantial experience in distillation and process design. Brendan holds a BS in chemical engineering from Columbia University.

137b Enhanced Performance of Thin Film Composite Forward Osmosis Membranes with Green Synthesized Graphene Quantum Dots

Haleema Saleem, Ahmad Fauzi Ismail, Peisean Goh, Syed Zaidi

The main drawbacks in forward osmosis (FO) process include high specific reverse solute flux (SRSF), reduced water flux, as well as fouling of membranes. In the current study, these challenges were addressed by the addition of graphene quantum dots (GQDs) obtained from a green sustainable source, in the selective layer of the thin film composite membranes, which were supported with an innovative polyethersulfone nanofiber support layer. Eucalyptus tree leaves were used to prepare GQDs by following a simple hydrothermal technique which required just distilled water, and there was no requirement for reducing agents or any organic solvents. The advanced solution blow spinning technology was employed for the preparation of the nanofiber support. The GQDs were incorporated in the selective later during the interfacial polymerization process to make thin film nanocomposite membrane and different characterizations were performed to confirm the formation of GQDs as well as its successful incorporation within the polyamide selective layer. These novel membranes have been tested for their performance in the FO process. To the best of our knowledge, this is the initial research work that reported the ultra high water flux as well as antifouling ability of the solution blown spun nanofiber-supported thin-film composite membranes. The effect of different GQDs concentrations on the performance of forward osmosis membrane, its long-term FO testing, effectiveness in cleaning, as well as its organic fouling resistance were examined. The performance of the thin film composite membrane in FO process was enhanced with the incorporation of 0.05 wt.% concentration of GQDs. Overall, the current work established that these advanced membranes showed a super high water flux of 3526 L/m^2h, low SRSF of 7.46 g/m^2h, better salt rejection as well as better antifouling performance in the forward osmosis process. These membranes have high potential for the scale-up of FO process.

137c Purification and Reuse of Poultry Slaughterhouse Wastewater with Membrane Technology

Faryal Fatima, Hongbo Du, Raghava Kommalapati

The conventional wastewater treatment methods used at poultry slaughterhouses which produce large amounts of wastewater face many challenges, especially the inability of recovering valuable nutrients and reusing treated water. In the last few years several researchers turn to membrane technology due to its enormous advantages over conventional methods. In this research a sequential membrane process of ultrafiltration (UF), forward Osmosis (FO) and reverse osmosis (RO) is proposed to treat semi-processed poultry slaughterhouse wastewater (PSWW) and to recover and reuse the water. The pretreatment of PSWW with UF removed 36.7% of chemical oxygen demand (COD), 38.9% of total phosphorous (TP), 24.7% of total solids (TS), 14.5% of total volatile solids (TVS), 27.3% of total fixed solids (TFS) and 12.1% of total nitrogen (TN). The PSWW was treated with FO membrane in FO mode, pressure retarded osmosis (PRO) mode, and L-DOPA coated membrane in the PRO mode. All these modes achieved complete removal of COD and TP. However, the FO mode showed the highest pollutant removal efficiency; 90.5% of TS, 85.3% of TVS, 92.1% of TFS, and 37.2% of TN. The FO mode achieved the highest average flux 10.4±0.2 L/m2-h. Since the FO mode showed the highest removal efficiency of pollutants and highest average flux, the FO mode was optimal for treating PSWW. The FO mode permeate was further treated by RO membrane to separate the filtrate from the draw solution and purify the PSWW for reusing it in the poultry industry. The RO significantly removed the COD, TS, TVS, TFS, and TP. However, the TN was reduced by only 62% because of the high ammonia concentration present in the draw solution. Overall, the sequential membrane process (UF-FO-RO) removed most of the pollutants from PSWW and produced water that could be reused for industrial poultry processing purposes.

137d Design Considerations for Chlor-Alkali Plants

Jonathan Cooper

Chlor-Alkali process systems require careful design consideration, due to the high temperatures and corrosive properties of the fluids that pass through them. Many of these corrosive fluids are incompatible with metal piping systems, which is why Chlor-Alkali facilities rely on thermoplastics, dual laminates, lined pipe and exotic alloys. Attend this interactive session to ensure a safe, effective piping system.

38a Ethanol to Jet Fuel Technology

Eseoghene Jeroro, Leigh Abrams, Manuela Serban

From blending mandates to tax penalties for conventional fuels, incentives across the European Union and the United States and other nations are increasing the demand for SAF, which is expected to continue to increase significantly through 2050.

Honeywell UOP is a leader in providing technology to produce renewable fuels from a variety of modern feedstocks that do not compete with food production or contribute to deforestation. Our Ecofining^{TM} technology, commercialized in 2013, has enabled the significant capacity of renewable diesel that meet summer and winter specifications, and SAF product that meets ASTMD7566 specifications for Synthetic Paraffinic Kerosene (SPK) from Hydroprocessed Esters and Fatty Acids (HEFA). However, existing technologies for HEFA-SPK production alone will not to satisfy the needs of the large and growing aviation market.

This presentation will cover the latest addition to Honeywell UOP's SAF portfolio – the new ethanol to jet fuel technology. Ethanol is an abundant, low-cost bio-feedstock which can be converted to sustainable aviation fuel (SAF). Depending on the type of ethanol feedstock used, jet fuel produced from Honeywell's ETJ process can reduce greenhouse gas (GHG) emissions by 80% on a total lifecycle basis, compared to petroleum-based jet fuel.[1] Our innovative Ethanol-to-Jet (ETJ) process also leverages more than a decade of experience and UOP's proven innovation in the space offering:

1. High jet yield output
2. Low CAPEX & OPEX
3. Reduced GHG emissions
4. Ability to commercialize at scale

38c Avoidable CO2 Emissions in Oil Refineries and the Role of Energy Efficiency in Meeting Net Zero Targets: The Case for the Crude Distillation Unit.

Emilio Diaz Bejarano, Francesco Coletti

Crude distillation units are generally known to be the largest energy consumer hence the largest contributor to CO2 emissions among all the processes in petroleum refining. As a result, this process is typically the main target to apply energy efficiency and CO2 reduction programs. These may include improved energy integration, management of operations and fouling mitigation.

In this presentation, we will address three aspects which are key to understand the potential benefits of applying technologies that reduce CO2 emissions in CDUs and for which only sparse or outdated information are available in the public domain. In particular we will assess:

1. The contribution of CDUs to total refinery emissions, which depends on i) the complexity of the refinery; ii) the efficiency of furnaces and the design of the preheat trains (energy integration) and iii) the severity of fouling and its management.
2. The fraction of the CDU emissions that are due to fouling.
3. The fraction of CDU emission that is avoidable by improved management of fouling in the pre-heat train.

For the first point we will present estimations are based on publicly available refinery emissions, assumptions based on the European CO2 allowance system and Hexxcell's own research. For the latter two, evaluations will be presented based on actual plant data from 10 refineries monitored on a regular basis. Results are then compared against the few previous studies available in the literature.

AUTHOR INDEX